YOUR KNOWLEDGE HAS VALUE

- We will publish your bachelor's and
  master's thesis, essays and papers

- Your own eBook and book -
  sold worldwide in all relevant shops

- Earn money with each sale

Upload your text at www.GRIN.com
and publish for free

Nora Görne

The Influence of the Tourist Industry and Municipality on Cultural Changes

GRIN Publishing

Bibliographic information published by the German National Library:

The German National Library lists this publication in the National Bibliography; detailed bibliographic data are available on the Internet at http://dnb.dnb.de .

Imprint:

Copyright © 2010 GRIN Verlag, Open Publishing GmbH
Print and binding: Books on Demand GmbH, Norderstedt Germany
ISBN: 978-3-640-99563-9

Nora Görne

UK English – MLA Style

23 November 2010

The Influence of the Tourist Industry and Municipality on Cultural Changes

That tourism flows pollute the local culture[1] and degrade the authenticity of a location (Desforges 517-518) has almost become a truism. In fact, as Coleman and Crang observe such critiques are as old as tourism itself (qtd. in Desforges 522-524) and are linked up with a wrong understanding of places – at least from a human geographical point of view. If detractors state that a locality has become "inauthentic" under touristic influence than in reverse it must have been "authentic", meaning genuine, unaffected and stable, before travelers visited it and brought their own culture. In contrast there is basically no place which developed without the influence of travelers. As Doreen Massey notes a place is not the way it is because of internal characteristics but because of global linkages (qtd. in Crang 45-49) and as Desforges adds "tourism and travel are merely a new form of interconnection between places" (523). For instance the Dutch city Middelburg gained some of its today's distinctive appearance and character by the slave trade and the resulting prosperity some hundred years ago (Encyclopædia Britannica). As Greenwood concludes "some of what we see as destruction is construction" (182). However, distinctive local traditions like certain ceremonies can change or even lose not only their original meaning but also the sense for residents under the influence of mass tourism (Hall and Page 122; Greenwood 181-182). One example for such a change of meaning can be seen in the development of the wine festivals in the Palatinate in Germany which will be examined later.

In academics, one of the main critiques of tourism is coined by Davydd J. Greenwood who argues that local culture gets commodified under touristic influence. As a certain amount of strangers attend an activity, like a ritual or celebration of cultural or historical signifiance for the local, "without their consent" and without "reimbursing them" their culture becomes a commodity and eventually often ends up as a tourist attraction but "meaningsless to the people who once believed in it" (173) and only gets preserved to make money. Instancing the

---

1   For the scope of this essay it is enough to define culture according to Pestalozzi as everything which is "typical for a human community in a specific region" (qtd. in Thiem 27)

"Alarde", a huge annual parade commemorating the resitance of the Basque city Fuenterrabia against the sixty-nine days lasting French siege in 1638, Greenwood shows how an internal ritual was transformed into a public touristic attraction within two years with the aid of municipal promotion. In the former times the Alarde was was not only a re-enactment of a historical event but also mattered as a display of their Basque identity. Then in the course of an increased publicity for the city when a hotel was opened in the rebuild fortress in 1969 the Alarde was added to the "basic tourism package" (177). The municipality tried to change the way this ritual was performed, so that hotel guests could see it, and even wanted to show it twice a year. The initial bewilderment of the inhabitants soon changed in a renunciative stance and the municipality encountered problems finding participants for this event. The prior ritual which was seen as a "performance for the participants" (176) was transformed into a show, a tourist feature.

This example shows quite well what Greenwood means with his term "cultural commoditization." Some people would consider this also as a prime example for a case in which tourism, or rather the "capitalist development" (Mowforth and Munt, 40), spoiled or destroyed the local culture. For the contemporary residents, this might be true because they lost an internal ritual which was once meaningful to them. However, whether a certain change of a ritual is positive or negative should only be judged from the perspective of the residents – even if they might also tend to disagree to cultural changes too quickly. Criticism by outsiders is therefore often connected with a "scarce empathy" (Thiem 28) with the locals. Furthermore, criticism sometimes originates in a "cultural pessimism" which values every variation from their imaginative, romantic picture of a certain place as a step towards cultural decay (Thiem 28). Then we also have to think about the impact that the tourism has on the economic status of the town Fuenterrabia (or any other touristic city) which might not be insignificant on the long run. When the tourist industry grows, foreign people will move to this place on their search for a job and will bring their own culture and innovations. Another point is the creation of cultural and sporting events for tourists from which residents could profit, too. All these factors would most likely be influential on the construction of the culture in the future. Already with the help of this small contemplation we can see that it is basically out of question that tourism simply destroys the local culture.

Nevertheless we should not forget the dangers which entail tourism and can influence the culture negativly. We have already examined two of them: "commercialisation" and the "modification of nature of [an] event" (Hall 122; Lea 73). Other points are the creation or intensification of stereotypes which eventually could change the "perception of [a] cultural identity by both locals and tourists" (Hall 131) and can be observed by looking in a random Dutch souvenir shop – which does not have wooden or at least stylised manufactured "klompen"? These clogs, like tulips and windmills, have become a symbol for the Dutch even if they are not worn anymore in the Netherlands. Refering back to the economical aspects, a big amount of new jobs created for a growing tourist industry in a certain place can also lead to uncertainty in the sedentary population. Thiem explicates that the values of the existing jobs might change, prices may rise and a rivalry between tourism and agricultural concerning the land is possible. The latter reason could even produce renunciation, apathy or xenophobia when residents feel that they have to subordinate their interests under the ones by the tourists or the tourist industry (30). The same applies if locals get the impression that they lose their cultural identity or that certain traits of their own culture get estranged (29).

This is what happened to some extent to the Palatine wine festivals, specifically to the "Dürkheimer Wurstmarkt". This event is the biggest wine festival of the world with over 600.000 visitors in the nine days it takes place (Dnews). The Wurstmarkt, which literally means "sausage market", has a long history and evolved originally from medieval pilgrimages to a nearby chapel. The resulting "Michaelismarkt" which was firstly mentioned in 1417 was later officially transformed into a Kermesse ("Kirchweihfest"). The wine festival gained rapidly in importance and was joint by several amusement stands and a circus at the beginning of the 19th century and by some ride attractions at the middle of the 20th century ("Dürkheimer Wurstmarkt"). Today it is basically a fun fair for everyone focused on rides rather than a wine festival for the locals.

It is impossible to estimate the proportional average of tourists and locals on the Wurstmarkt in comparison to other wine festivals. However, I found it striking that during a small survey I did with six acquaintances which come from the region round Bad Dürkheim all agreed that the Wurstmarkt is rather a crowded tourist attraction which they cannot identify as a part of the local tradition of wine festivals. Five of them prefer going to smaller wine festivals which have not been promoted extensivly in tourist leaflets. One respondent mentioned the reason that he sees wine festivals as a possibility to meet friends and other *local* inhabitants and enjoy together the young wine. Another interviewee criticised the homogenization of wine festivals remarking that stands are getting more and more similar as tourism spreads and it is hard to find a village which offers something "original." Besides, it is noticeable that the distinction between wine festivals, kermessen, wine tithes and similar vanishes, sometimes unintentional but sometimes also implied by advertiser for a better promotion. The "almond blossom festival" of Gimmeldingen has originally nothing to do with wine but today it heralds the wine festival season normally in March (Sebastian). In contrast, in the early 50s the first festivals started in July and ended in October (Landesarchivverwaltung Rheinland-Pfalz). It seems that the culture is much more "spoiled" by the side effects of tourism – for example the municipal public relations and advertising agencies – than by the tourists itself. Their promotion contributes to a large part to the negative effects of tourism judging by the locals as we have already seen in Greenwood's example.

In Hawai'i we find an example how the tourist industry transformed and even exploited local culture for their own purposes: The little islands of Hawai'i are popular as a wedding place for both North Americans and Japanese with "over 46,000 tourist couples […] in 2000" (McDonald 174). Both nationalities are looking for an "exotic experience" though the former are offered romantic weddings on the beach, embodying an "oriental fantasy", and the latter wish for a simplified Western-style ceremony ("chapel wedding") expressing an "occidental" dream (McDonald 176). Actually, the concept of Hawaiian weddings and the tourist's image of islands as a "tropical paradise" (McDonald 176) has been highly influenced by an American hotelier, Grace Guslander, who set with her "Coco Palms" a standard for further resorts. The way she mixed local culture with other influences (TenBruggencate; McDonald 176) trying to meet the oblique expectations of Westerners exemplified the concept of creating "artifically" a culture.

This especially becomes clear when looking at the way the ceremonies were customised to Japanese couples who arrived in the 1990s (McDonald 177). They dream of a Western-style ceremony with a typical white dress for the bride and representative Christian symbols. Weddings are therefore "staged" in local chapels. As the number of Japanese tourists has been rising steadily, in 2000 30.000 Japanese couples married in Hawai'i (McDonald 174), firms like "Watabe Wedding" struggled with finding new wedding venues. They ended up in conducting "six to eight one-hour weddings per day" (McDonald 190-191) in one chapel, building new ones and using restaurants, hotels and homes for ceremonies. This has lead to protests among the locals (McDonald 181, 183, 191); nevertheless, the state Hawai'i Visitors and Convention Bureau even included weddings in their "Aloha Magic" campaign (qtd. in McDonald 181) effecting even more tourists.

One could argue now that the extensive bridal tourism is becoming a new part of Hawaiian culture: At least with the new hotels, restaurants and chapels, which are build for the tourists, it completely transforms the former spatial pattern and look of the islands and

accounts, according to our definition of culture[2], to *new* characteristics of the Hawaiian.

However, with that implication we could embrace every cultural change affected by tourism

as positive, a point Greenwood is alerting, too (182). As I already mentioned, it basically

depends on the definition of "damaging" - when is a culture not "authentic" anymore or

"spoiled"? In my opinion, that is the case when locals feel alienated in their own space. When

the changes of local culture were not affected by tourism itself but implied by municipality

like in Fuenterrabia and in the Palatinate or even created by the tourist industry like we have

seen it in Hawai'i. Greenwood additionally alludes that more research has to been done

concerning "communities as a complex process of stability and change" before we can

include alternations affected by tourism in the conceptualisation (182).

---

2   For the scope of this essay it is enough to define culture according to Pestalozzi as everything which is
    "typical for a human community in a specific region" (qtd. in Thiem 27)

Works Cited

Crang, Phil. "Local – Global." *Introducing Human Geographies.* Ed. Paul Cloke, Phil Crang

and Mark Goodwin. London: Routledge, n.d. 34-49. Print.

Desforges, Luke. "Travel and Tourism." *Introducing Human Geographies.* Ed. Paul Cloke,

Phil Crang and Mark Goodwin. London: Routledge, n.d. 517-526. Print.

Dürkheimer Wurstmarkt. Stadtverwaltung Bad Dürkheim, n.d. Web. 18 Nov. 2010.

Greenwood, Davydd J. "Culture by the Pound: An Anthropological Perspective on Tourism

as Cultural Commoditization." *Hosts and Guests: The Anthropology of Tourism.* Ed.

Valene L. Smith. 2nd ed. Philadelphia, Pennsylvania: University of Pennsylvania

Press, 1989. Print.

Hall, Colin Michael, and John Page. *The Geography of Tourism and Recreation.* London:

Routledge, 1999. Print.

Lea, John. *Tourism and Development in the Third World.* London: Routledge, 1988. Print.

McDonald, Mary G. "Tourist Weddings in Hawai'i: Consuming the Destination." *Seductions

of Place.* Carolyn Cartier and Alan A. Lew, eds. New York: Routledge, 2005. 171-

192. Print.

"Middelburg." *Encyclopædia Britannica.* 2010. Encyclopædia Britannica Online. 18 Nov.

2010.

Mowforth, Martin and Ian Munt. *Tourism and Sustainability: Development, Globalisation

and New Tourism in the Third World.* Oxford: Taylor & Francis, 2008. *Google Books.*

Web. 23 Nov. 2010.

Sebastian, Heike. "Das Gimmeldinger Mandelblütenfest." *suite101.* Suite101.com Media Inc,

6 Feb. 2010. Web. 23 Nov. 2010.

TenBruggencate, Jan. "Grace Guslander, Visionary Hotelier, Dies." *The Honolulu Advertiser.*

Cisan Online Ventures, 6 Apr. 2000. Web. 23 Nov. 2010.

Thiem, Marion. "Tourismus und kulturelle Identität." *Tourismus und kulturelle Identität –*

Die Bedeutung des Tourismus für die Kultur touristischer Ziel- und Quellgebiete.

Bern - Hamburg, 1994. *Bundeszentrale für politische Bildung.* Web. 18 Nov. 2010.

<http://www.bpb.de/files/JQIKZM.pdf>

"Vor 55 Jahren." *Landesarchivverwaltung Rheinland-Pfalz.* Landeshauptarchiv Koblenz, 7

Sep. 2007. Web. 18 Nov. 2010.

<http://www.landeshauptarchiv.de/index.php?id=320>

"Wurstmarkt lockt mit neuen Attraktionen." *Dnews.* Dnews.de, 1 Sep. 2010. Web. 18 Nov.

2010.

Word Count: 1944